BLU[illegible] NOTEBOOK

Architectural Masterpieces

This notebook belongs to

GLASGOW SCHOOL OF ART
Charles Rennie Mackintos

GLASGOW, SCOTLAND (1909)

VILLA SAVOYE PARIS

POISSY, PARIS, FRANCE (1929)

CHAPEL OF NOTRE DAME DU HAUT
Le Corbusier

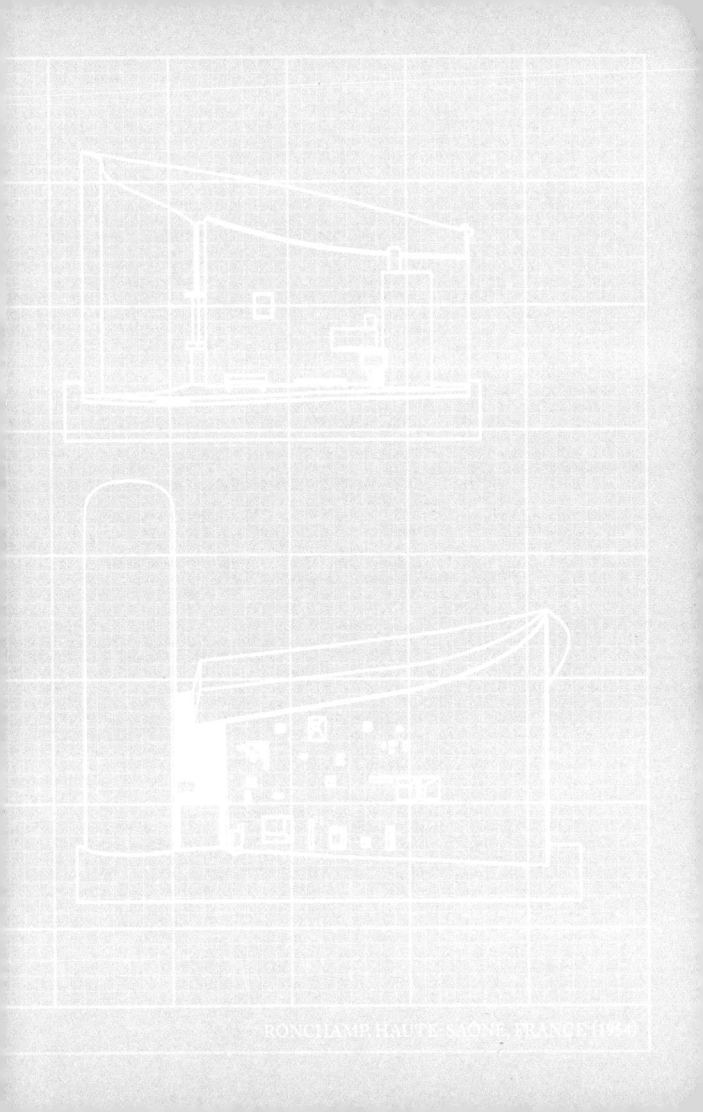
RONCHAMP, HAUTE-SAÔNE, FRANCE (1954)

BILBAO, SPAIN (1997)

HSBC
Norman Foster

HONG KONG, CHINA (1985)

SYDNEY OPERA HOUSE
Jørn Utzon

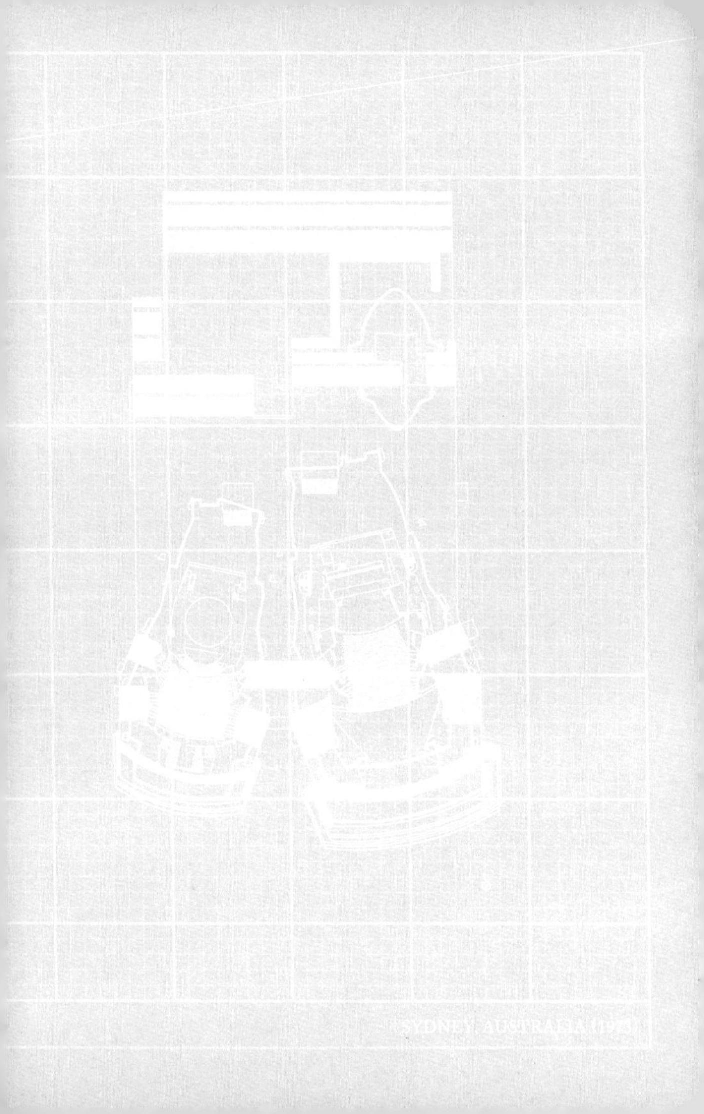
SYDNEY, AUSTRALIA (1973)

BAUHAUS

Walter Gropius

DESSAU, GERMANY (1926)

SENDAI MEDIATHEQUE

TUGENDHAT VILLA

Ludwig Mies van der Roh

BRNO, CZECH REPUBLIC

RIETVELD SCHRODER HOUSE
Gerrit Thomas Rietveld

UTRECHT, THE NETHERLANDS | 1924

WEISHAUPT FORUM
Richard Meier

SCHWEDT, GERMANY [illegible]

EXETER LIBRARY
Louis Kahn

EXETER, NEW HAMPSHIRE, UNITED STATES 1972

SÄYNÄTSALO TOWN HALL
Alvar Aalto

SÄYNÄTSALO, FINLAND (1952)

HOUSE IN CASCAIS
Eduardo Souto de Moura

CASCAIS, PORTUGAL (2002)

THE HILL OF THE BUDDHA
Tadao Ando

SAPPORO, JAPAN (1982)

NITEROI CONTEMPORARY ART MUSEUM

RIO DE JANEIRO, BRAZIL (1996)

WILLEMSPARK SCHOOL
Herman Hertzberger

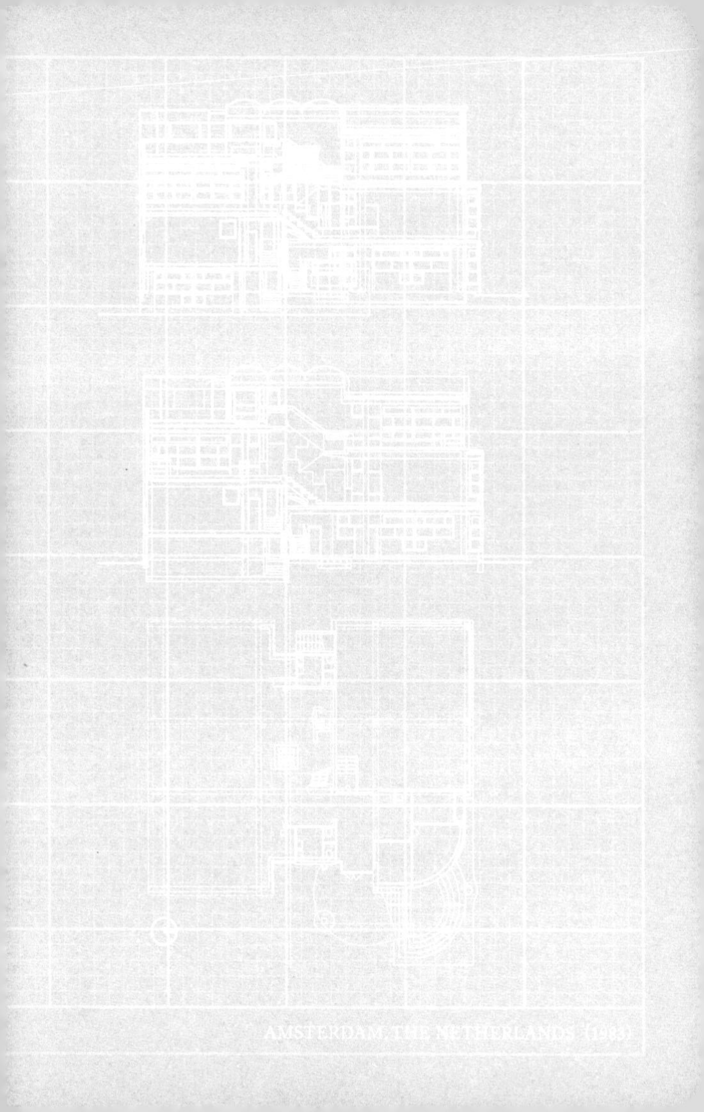
AMSTERDAM, THE NETHERLANDS (1983)

LA CASA MILÀ
Antoni Gaud

BARCELONA, SPAIN (1910)

WILLIS FABER & DUMAS HEADQUARTERS
Norman Foster

IPSWICH, GREAT BRITAIN 1975

BURGERWEESHUIS
Aldo van Eyck

AMSTERDAM, THE NETHERLANDS (1960)

CENTRAAL BEHEER OFFICE BUILDING
Herman Hertzberger

APELDOORN, THE NETHERLANDS (1972)

ROBIE HOUSE
Frank Lloyd Wright

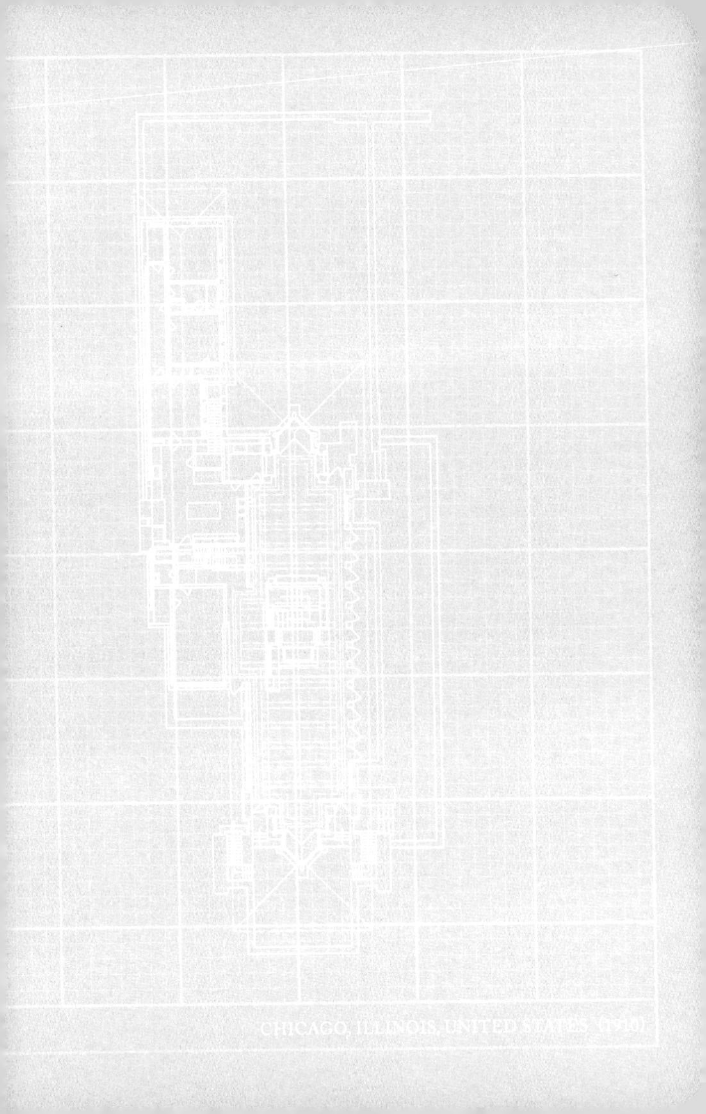
CHICAGO, ILLINOIS, UNITED STATES (1910)

GOTHENBURG CITY HALL
Gunnar Asplund

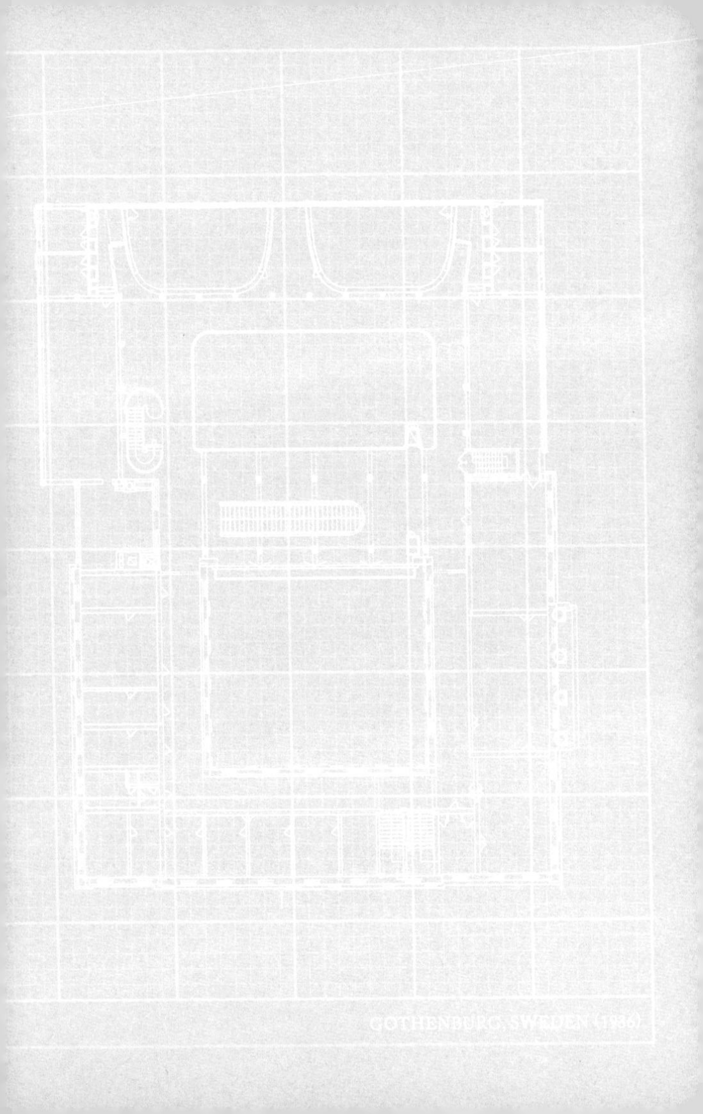
GOTHENBURG, SWEDEN (1986)

HALEN ESTATE
Atelier 5

BERNE, SWITZERLAND (1961)

BERLINER PHILHARMONIE
Hans Scharoun

JÜDISCHES MUSEUM
Daniel Libeskind

BERLIN, GERMANY (2001)

NEW YORK, NEW YORK, UNITED STATES (U.S.)

SUBANG JAYA, SELANGOR, MALAYSIA (1992)

CAMBRIDGE, UNITED STATES (1946)

PALACE OF THE ASSEMBLY
Le Corbusier

CHANDIGARH, INDIA

PALALOTTOMATICA
Marcello Piacentini, Pier Luigi Nervi

ROME, ITALY

QUERINI STAMPALIA FOUNDATION

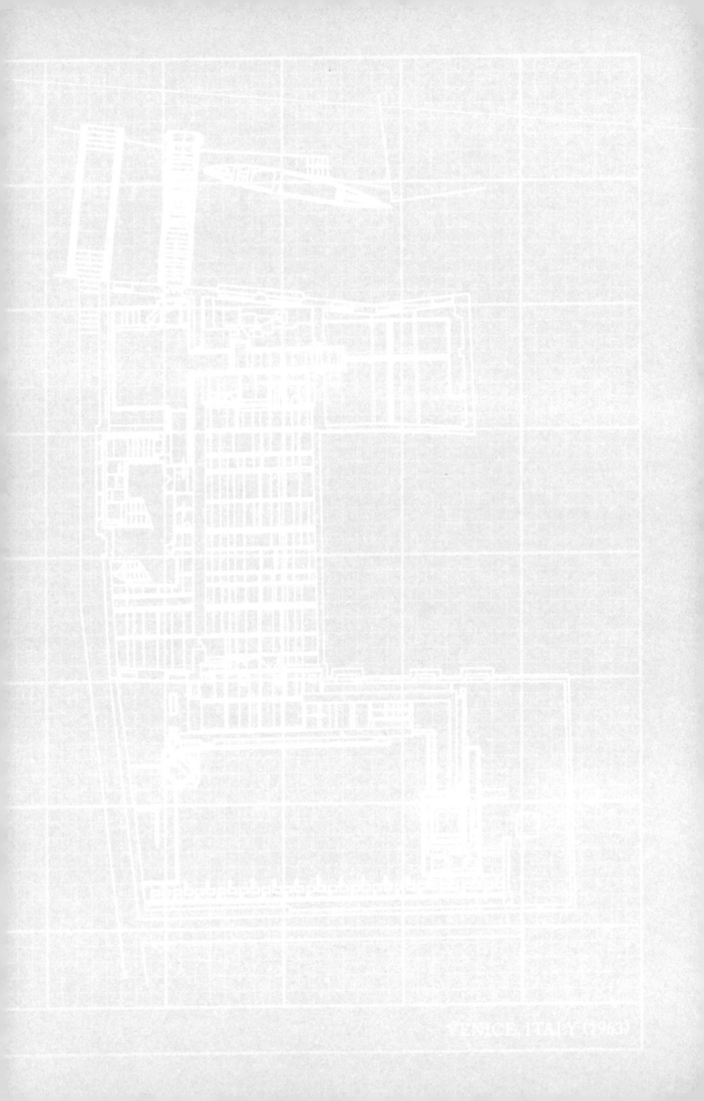
VENICE, ITALY

RICHARDS MEDICAL RESEARCH LABORATORIES
Louis Kah

PHILADELPHIA, PENNSYLVANIA, UNITED STATES (1961)

SC JOHNSON ADMINISTRATION BUILDING AND RESEARCH TOWER

Frank Lloyd Wright RACINE, WISCONSIN, UNITED STATES (1950)

SEAGRAM BUILDING
Mies van der Rohe & Philip Johns

NEW YORK, NEW YORK, UNITED STATES (1958)

SEATTLE, WASHINGTON, UNITED STATES (1997)

STOCKHOLM PUBLIC LIBRARY
Gunnar Asplund

STOCKHOLM, SWEDEN (1928)

STOCLET PALACE
Josef Hoffmann

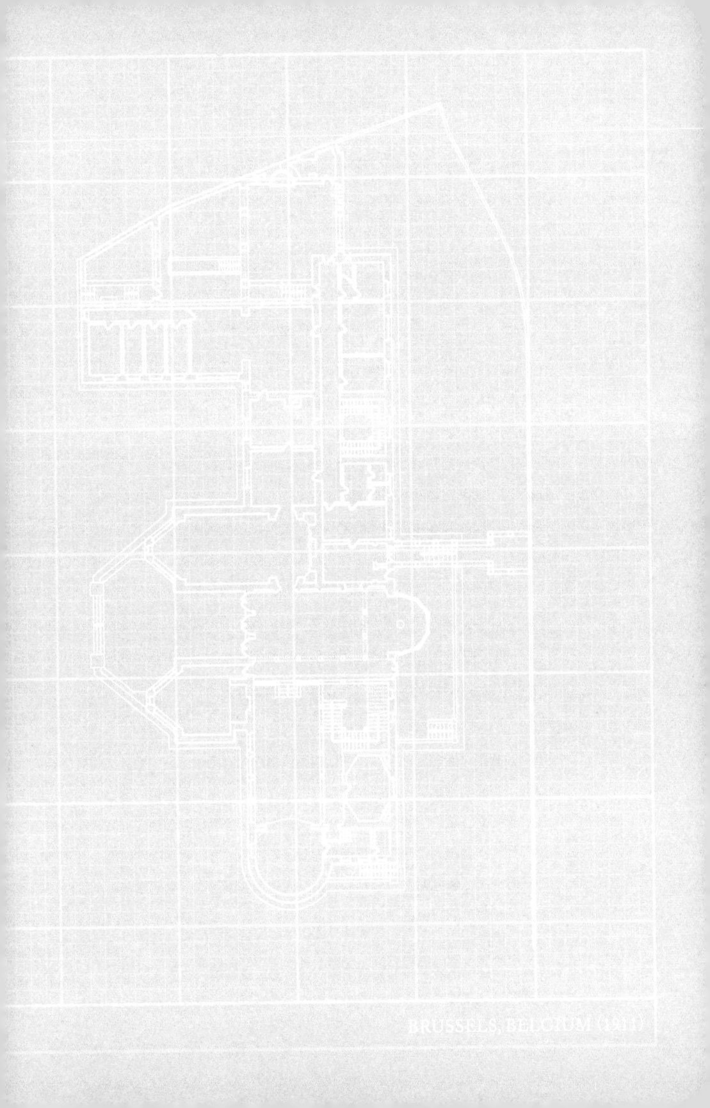
BRUSSELS, BELGIUM (1911)

THE ECONOMIST BUILDING
Alison & Peter Smithson

LONDON, GREAT BRITAIN (1964)

THE FORD FOUNDATION

Kevin Roche John Dinkeloo & Associates

NEW YORK, NEW YORK

VILLA MAIREA
Alvar Aalto

NOORMARKKU, FINLAND (1939)

PARIS FRANCE ([illegible])

AUDITORIUM PARCO DELLA MUSICA
Renzo Piano

PARIS, FRANCE (1979)

PALACE OF THE ASSEMBLY
Le Corbusier

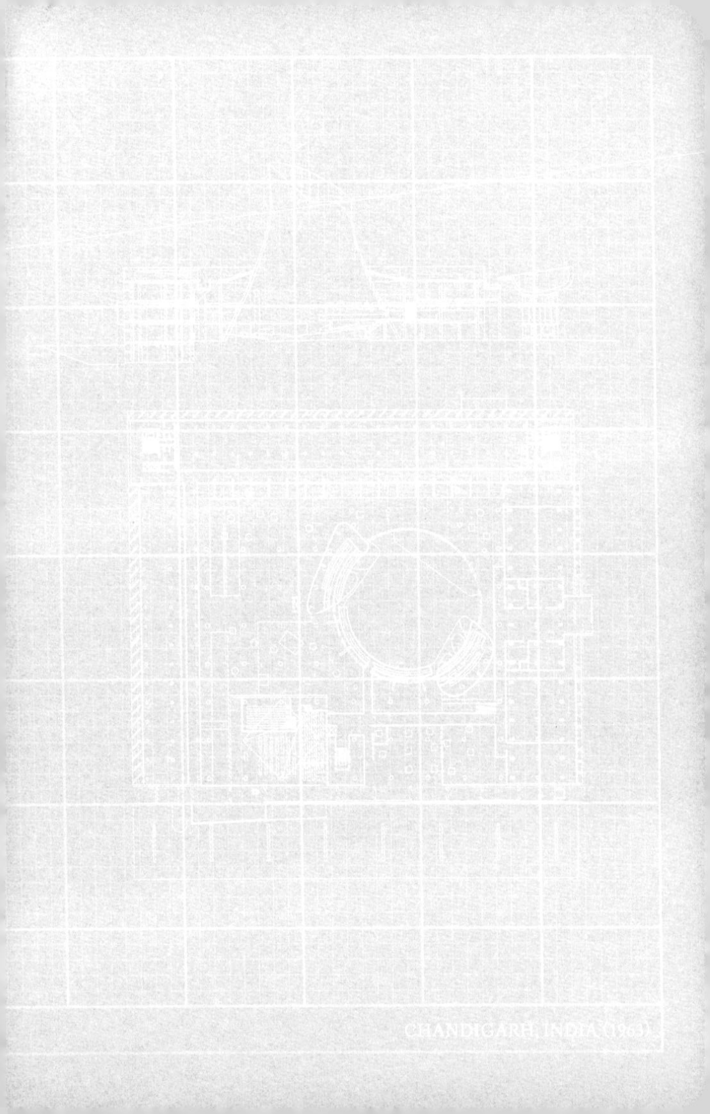
CHANDIGARH, INDIA

CENTRO GALEGO DE ARTE CONTEMPORANEA
Álvaro Siza Vieira

SANTIAGO DE COMPOSTELA, SPAIN (1993)

NEUE STAATSGALERIE
James Stirling